Animal Grief

HOW ANIMALS MOURN

DAVID ALDERTON

Hubble & Hattie

The Hubble & Hattie imprint was launched in 2009 and is named in memory of two very special Westies owned by Veloce's proprietors.
Since the first book, many more have been added to the list, all with the same underlying objective: to be of real benefit to the species they cover, at the same time promoting compassion, understanding and co-operation between all animals (including human ones!)
Hubble & Hattie is the home of a range of books that cover all-things animal, produced to the same high quality of content and presentation as our motoring books, and offering the same great value for money.

MORE TITLES FROM HUBBLE & HATTIE

A Dog's Dinner (Paton-Ayre)
Animal Grief: How animals mourn (Alderton)
Cat Speak (Rauth-Widmann)
Clever dog! Life lessons from the world's most successful animal (O'Meara)
Complete Dog Massage Manual, The – Gentle Dog Care (Robertson)
Dieting with my dog (Frezon)
Dog Cookies (Schöps)
Dog-friendly Gardening (Bush)
Dog Games – stimulating play to entertain your dog and you (Blenski)
Dog Speak (Blenski)
Dogs on Wheels (Mort)
Emergency First Aid for dogs (Bucksch)
Exercising your puppy: a gentle & natural approach – Gentle Dog Care (Robertson & Pope)
Fun and Games for Cats (Seidl)
Know Your Dog – The guide to a beautiful relationship (Birmelin)
Miaow! Cats are ncer than people (Moore)

My dog has arthritis – but lives life to the full! (Carrick)
My dog is blind – but lives life to the full! (Horsky)
My dog is deaf – but lives life to the full! (Willms)
My dog has hip dysplasia – but lives life to the full! (Haüsler)
My dog has cruciate ligament injury – but lives life to the full! (Haüsler)
Older Dog, Living with an – Gentle Dog Care (Alderton & Hall)
Partners – everyday working dogs being heros every day (Walton)
Smellorama – nose games for dogs (Theby)
Swim to recovery: canine hydrotherapy healing – Gentle Dog Care (Wong)
The Truth about Wolves and Dogs (Shelbourne)
Waggy Tails & Wheelchairs (Epp)
Walking the dog: motorway walks for drivers & dogs (Rees)
Walking the dog: walks in France for drivers & dogs (Rees)
Winston ... the dog who changed my life (Klute)
You and Your Border Terrier – The Essential Guide (Alderton)
You and Your Cockapoo – The Essential Guide (Alderton)

Photo credits
The illustrations used in this book were kindly supplied by the following photographers through www.shutterstock.com, except where indicated by an asterisk:-

5t JKlingebiel 5b Trevor kelly 6l Panos Karapanagiotis 6r Panos Karapanagiotis 7 Sue Smith 8l Colin Edwards Photography 8r Michael Avory 9 eans 10 Raedwald 11 Mike Tan C T 12 john austin 13bl vitor costa 13t Xavier MARCHANT 14 First Class Photos PTY LTD 15 Graham Taylor 16 Benjamin Albiach Galan 17 Alila Sao Mai 18-19 A v d Wolde 20 Jeffrey M Frank 21 Similaun Man 23 Johan Swanepoel 24 Petr Jilek 25 Patryk Kosmider 26 Mogens Trolle 27 Tiago Jorge da Silva Estima 28 Four Oaks 29t n/a 29b Nickolay Stanev 30 Jeff Schultes 31l Stubblefield Photography 31r Borhuah Chen 32 Eric Gevaert 33 palko72 34 Elizabeth Dover 35 SHADOWMAC 36 JKlingebiel 37 Ronald van der Beek 38-39 Gentoo Multimedia Ltd 40tl idreamphoto 40r melissaf84 43 Tatonka 44 Knumina 45 Dale Mitchell 46l SouWest Photography 46-47 Rich Carey 49tl Nephron* 49tr iDesign 49b Graeme Shannon 50 Hugh Lansdown 51t Four Oaks 51b John C Hooten 52 Four Oaks 53 Four Oaks 55 Shamleen 56l FunkMonk* 56r Dave McAleavy 57 Ronald van der Beek 58 Jirsak 59t Achim Baque 59bl Graham Bloomfield 59br Angelo Giampiccolo 61 Sandra Cunningham 62l Rob kemp 62r John Carnemolia 63 Botond Horváth 64 Muramasa* 65 Heartland 66 BMCL 67 Montanabw* 68 n/a 69 Michael Pettigrew 70 Sari ONeal 71 Michelle D Milliman 72 RamonaS 73 Michael Klenetsky 74l Lindsey Eltinge 74r worldswildlifewonders 75 trucic 76bl Studio 37 76tr Tischenko Irina 77 Eric Isselée 78 guentermanaus.

Front cover JKlingebiel; spine Gastón M Charles; back cover left Henk Bentlage; tc Laszlo Szirtesi; tr xJJx; bc Marcel Jancovic; br tobe_dw; inside front flap & 80 Eric Isselée; front endpaper Monica Szczupider/National Geographic My Shot/National Geographic Stock; rear endpaper JKlingebiel;

WWW.HUBBLEANDHATTIE.COM

First published in October 2011 by Veloce Publishing Limited, Veloce House, Parkway Farm Business Park, Middle Farm Way, Poundbury, Dorchester, Dorset, DT1 3AR, England. Fax 01305 250479/e-mail info@ hubbleandhattie.com/web www.hubbleandhattie.com
ISBN: 978-1-845842-88-8 UPC: 6-36847-04388-2 © David Alderton and Veloce Publishing 2011. All rights reserved. With the exception of quoting brief passages for the purpose of review, no part of this publication may be recorded, reproduced or transmitted by any means, including photocopying, without the written permission of Veloce Publishing Ltd. Throughout this book logos, model names and designations, etc, have been used for the purposes of identification, illustration and decoration. Such names are the property of the trademark holder as this is not an official publication. Readers with ideas for books about animals, or animal-related topics, are invited to write to the editorial director of Veloce Publishing at the above address.
British Library Cataloguing in Publication Data – A catalogue record for this book is available from the British Library. Typesetting, design and page make-up all by Veloce Publishing Ltd on Apple Mac. Printed in India by Replika Press

Contents

Changing times, changing ways

Emotions in animals remain a controversial subject. Some people do not believe that animals are capable of feeling emotion generally, let alone grief, possibly because grief represents one of the most complex of emotions. It is acquired with age, rather than being innate like anger, and is not often displayed.

Many owners would privately accept that their pets do show signs of emotion, including grief, but these have been essentially anecdotal accounts, which have not been followed up. Today, though, thanks in part to the internet, there is much greater debate surrounding this topic, which has now triggered serious scientific interest.

Differing thoughts about the way in which animals perceive the world extend back to the dawn of history. These have been markedly influenced by the way in which animals have been perceived in different cultures. Unfortunately, our beliefs about emotions in animals have been clouded over the course of many centuries by theological dogma, and issues surrounding the human condition and soul. This has had the effect of blocking genuine investigation into this field.

A radical rethink of the concept of animal emotions is currently taking place today, however, inspired by the way in which science is providing some remarkable insights into animal behaviour. These findings are causing us to view the natural world in a completely different way.

Crocodilians, for example, have emerged as devoted parents, rather than mechanistic killers. Elephants can keep in touch with each other across long distances by using infrasound, which is inaudible to our ears. The song of whales reverberates through the oceans for the same purpose, albeit increasingly drowned out by interference from ships' sonar.

THE CONCEPT OF THE SOUL

When it comes to discussions about whether or not animals can feel grief, these have been closely bound up over centuries with the concept of animal consciousness. The Greek philosopher Aristole (384-322 BC) was one of the first people to consider the possibility of animals having souls. What made his approach different, however, as set out in his work *De Anima* (meaning *On the Soul*), was that he sought to consider this from a scientific rather than a theological perspective, unlike many later writers.

His conclusion was different, too, because he viewed the soul as the life force, rather than an entity that survived death, as encapsulated in subsequent Christian teaching, and that of other religions, too. Aristotle therefore considered that all life forms had

Grief is an emotion that is linked with social awareness.

souls, although he sought to distinguish between them.

Plants, in his view, had the least sophisticated form of soul, as evidenced by the fact that they did not move. The soul of plants allowed them simply to grow and reproduce. Animals, in contrast, possessed a higher form of soul, combining a vegetative soul with a rational soul, which allowed them to feel things and move. Those of humans were at the highest level of all, though, because people are capable of being rational and possess the ability to reason.

Yet even Aristotle considered that one of the reasons why animals had a lesser soul was because,

A young crocodile rests on his mother's back.

Animal Grief

A statue of Aristotle at his birthplace at Stageira, Greece. Not only did he develop a concept of the soul, he also laid the foundations for modern-day zoology.

A statue honouring the ancient Greek philosopher Plato, which can be seen at the Academy of Athens building in Athens.

unlike people, they were incapable of having feelings. In Aristotle's world, animals would not have been considered capable of feeling grief; this would have been the exclusive preserve of humans.

His approach was, nevertheless, quite radical, when set against established classical teachings up to that point. Plato had previously suggested that the soul was an independent entity which existed within the body, and survived death. Aristotle, however, was convinced that the soul represented life in the organism it inhabited, and could not survive death. He believed it was the soul which served to distinguish living creatures from inanimate objects.

Another point of distinction — which is still apparent in our culture today — is Aristotle's assertion that the location of the rational soul lay not in the human brain, but rather in the heart.

Aristotle developed the original scientific classification of animals, being the first to distinguish between what we now recognise as vertebrates and invertebrates. Alongside this, however, he also sought to create what he called a *scala naturae*, or 'great chain of being,' which ranked life forms into eleven different groups, from plants at the bottom of the chain up to man, followed by God at the top. It was to have a major influence on our view of the natural order right through to the mediaeval period, and well beyond, with its influence still being apparent up until the 1800s.

There is no doubt that Aristotle's work helped to lay the foundations for development of modern zoology, which is the branch of science covering the study of animals. He distinguished marine mammals from fish, for example, and also recognised that cartilaginous fish, such as rays and sharks, formed a separate group in their own right.

The issue of whether or not animals have a soul was subsequently taken up by St Thomas Aquinas (1225-1274), an Italian who was a member of the Dominican order, and one of the most influential theologians in the Catholic Church. He studied a number of Aristotle's works in detail, including *On the Soul*, and concurred with his view that man as a species was defined by having an animal body, which was illuminated by the possession of a rational soul. But a key point of

St Thomas Aquinas – a highly influential Christian thinker who lived during the Middle Ages.

distinction with Aristotle's thinking was that the soul of man survived death.

COMMUNICATION WITH ANIMALS

During the mediaeval period, one of the most famous of all religious figures, in the context of animals, was born in the early 1180s. Giovanni Francesco di Bernardone had a relatively short life, dying at about 45 years of age. During his life, though, he had some remarkable encounters with birds and animals, which seemingly reflected an unparalleled ability to communicate with them, and suggesting they had emotions.

On one occasion, he paused by the side of the road, heading off to preach to the birds that were flying around the neighbouring trees. His companions marvelled as they showed no fear, and did not fly off as he approached.

Another, more dangerous challenge for Giovanni occurred on the outskirts of the city of Gubbio, in the north-eastern part of the Italian province of Perugia in Umbria, where he was living at that time. A ferocious wolf was terrorising the local community, attacking people as well as livestock, and he decided to head into the surrounding hills in search of this feared predator.

Abandoned by his companions as he approached the wolf's lair, Giovanni made the sign of the cross when he encountered the animal, instructing him to desist in his attacks. Remarkably, according to contemporary accounts, the wolf then accompanied Giovanni back to the city where he had caused so much fear. Giovanni persuaded the townspeople that the wolf had abandoned his evil ways, and instructed them to provide food regularly. In addition, Giovanni agreed with the local dogs that they also would not harm the wolf.

An even more extraordinary event is documented as occurring right at the end of Giovanni's life. As he was dying, he expressed his thanks to his donkey, who had carried him for many years. When Giovanni passed away, those around recorded that the donkey

The burial place of St Francis d'Assisi, commanding an impressive view over the Umbrian countryside. He was interred here on 25 May, 1230.

Relatively few species were known in mediaeval times. Gorillas, for example, were not discovered until the 1800s.

cried for his much-loved owner, reflecting the grief of those who knew him.

These remarkable links with animals are just part of the reason why, shortly after his death, Giovanni was canonised on July 16, 1228 by Pope Gregory IX. Now universally known as St Francis of Assisi, he is recognised as the patron saint of animals.

AN OBSESSION WITH HIERARCHY

During the mediaeval period, animals were regarded as being less intelligent than people under the *scala naturae*, lacking in terms of mental attributes, and were not considered to possess a soul. They were ranked below man, followed by trees and plants, right down to minerals. Each hierarchy was divided into various sub-groups, extending to the King of Beasts in the case of animals, a description which is now generally considered to apply to lions, although elephants were also regarded as fulfilling this role back then.

One unusual aspect of this scheme, perhaps, was that wild animals such as lions were considered superior to domesticated animals, as they were more independent by nature, and could not be tamed. Bestiaries – illustrated compendiums featuring animals and plants, as well as minerals, in some cases — served to reinforce the religious order with moral messages.

Such publications drew heavily on early classical works, including Aristotle's *Historia Animalium*, for example, while some of their content represented a genuine advance in our understanding of the natural world, such as the concept of birds migrating, although this was not to be universally accepted for many centuries. In fact, not all of the animals included in such publications actually existed in reality, with the unicorn, of course, being a prime example of a fictional creature.

The lack of a soul and understanding meant that animals could not be considered sufficiently self-aware to express grief when one of their number died. The Age of Reason that followed saw a more secular view evolving, with philosophers such as Descartes (1596-1650) adopting what was essentially a mechanistic approach. Both men and animals were regarded as being like machines, but what now separated them was human consciousness, rather than a soul. This new belief, however, once again credited animals with being bereft of any emotional feelings such as grief.

THE CONCEPT OF ANIMISM

Aristole's belief in animals having a soul has been widespread in many cultures, and other religions, too. For example, it is a feature of faiths such as Hinduism, Buddhism, and Sikhism, and plays a central role in the culture of various indigenous peoples. In most cases, in line with Aristole's thinking, animal souls are considered inferior to human souls. However, Hinduism permits animal souls to move up into people through the process of reincarnation. It recognises two separate components: *dharma*, an instinctual animal element; and the free choice, which is what predominates in people, as reflected by *karma*.

A westernised form of belief that other creatures, aside from man, and sometimes inanimate objects possess a spirit or soul can be traced back to

observations by the German chemist Georg Ernest Stahl (1659-1734), which were published around 1720. He founded the cultural belief described today as animism, which derives from the Latin word for *anima*, meaning 'life' or 'soul.'

Stahl believed that the act of burning wood and reducing it to ash caused the 'vital force' — a form of soul which was present — to leave, as reflected by the fact that the ash weighed less than the original wood. The concept of animism was later revived by the anthropologist Sir Edward Tylor (1832-1917) in his book entitled *Primitive Culture,* which was published for the first time in 1871 and proved highly influential.

Tylor suggested the idea that other animals, and even inanimate objects, might possess souls was a view common in tribespeople around the world. This notion became popular amongst Tylor's fellow anthropologists, who felt this belief reflected a lack of cognitive understanding in what they considered to be primitive societies.

Today, however, the animistic view that man is part of nature and the natural world, rather than superior to it, would appear to be increasingly

The burning of wood gave rise to the concept of a 'vital force,' equivalent to the soul.

Animal Grief

accepted, especially given the environmental damage that more 'sophisticated' societies have inflicted on the planet since Tylor's pronouncements were originally published.

DARWIN'S IMPACT

It was the growth of interest in the emerging science of animal behaviour, which became known as ethology, that helps in part to explain why emotions in animals are now at last being considered seriously. Although best-remembered today as the founding figure of evolutionary thought, Charles Darwin (1809-1882) laid the foundations for the development of ethology as well, in his book *The Expression of Emotions in Man and Animals*. This was published in 1872, some 13 years after *On the Origin of Species*, which had rocked the scientific and religious communities to the core.

By the time that publication of his new work drew near, however, Darwin had wearied of the project,

A view of Christ's College, Cambridge, where Charles Darwin studied as a young man.

having had to revise the text extensively at proof stage. At the core of this book was his observation that there is a common link reflecting the ways in which both people and animals express their emotions through distinctive movements. This was in stark contrast to the existing belief, which had been derived essentially from the views of Sir Charles Bell (1774-1842) a Scottish surgeon, neurologist, anatomist, and philosophical theologian earlier in the nineteenth century.

Bell, seeking to reinforce the accepted differences between man and other animals, had decreed that people possessed a divinely created set of facial muscles in his publication *Anatomy and Philosophy of Expression*, which had been published in 1824. Darwin, however, showed a uniformity across the species in his work, and that man was *not* different, as Bell had claimed.

In its own way, therefore, *The Expression of Emotions in Man and Animals* was to be just as seminal as *On the Origin of Species*, driving forward debate, and ultimately helping to pave the way for what were to become new fields of scientific study, which have since flourished. Even Sigmund Freud (1856-1939), the founder of psychoanalysis, acknowledged a debt to Darwin for his analysis of emotions and behaviour.

Subsequently, behavioural science as a branch of zoology came to focus in the twentieth century on natural reactions, displayed by members of a species as a whole, rather than an individual. This gave rise to the creation of what became known as species ethograms, which record the main types of behaviour seen in species, and the frequency with which these are observed. Since then, more study of individual members of species has been carried out, assisting insights into emotions which may otherwise be missed.

TESTING SELF-AWARENESS

It was also Charles Darwin who helped to pioneer research into the concept of self-awareness, which is a key indicator as far as determining whether or not animals may feel grief when one of their companions

Facial signs of emotion can be seen in many species, particularly primates. Curiosity, for example, is very evident here in the faces of these young orang–utans.

dies. Darwin simply held up a mirror as an experiment when he visited what is now London Zoo, allowing Jenny, the orang-utan to see her reflection.

He then recorded her reactions, although he was unable to be certain whether or not the ape was simply playing, or trying to communicate by means of her expression with what she might have thought to be another animal. In the context of animal grief,

however, the outcome of this simple test is critical. In order to be affected by grief, it is clearly essential that an individual can be shown to be aware of its own existence, and that of others around it.

Remarkably, perhaps, it was not until the 1970s that Darwin's basic test was ultimately refined by Gordon G Gallup, Jr, a member of the University of Albany's Psychology Department in New York, for

Animal Grief

the purpose of defining self-awareness. He came up with the idea of using an odourless dye, two spots of which were applied to the subject's body: one would be located on an area of the body which would be clearly visible in the mirror, whilst the other would be hidden from view.

If the animal ignored the spot of dye that was out of sight of the mirror, whilst showing obvious signs of recognising the dye visible in the mirror, this could be interpreted as an indication that he was self-aware. He might, for example, turn his body slightly to get a better view of the dye, or may touch the area of his body where the dye had been applied, having seen it there in the mirror.

What is particularly interesting is that, at first, all animals which are exposed to a mirror behave rather like Darwin's orang-utan, believing the image to be that of another member of their own kind. Even more remarkably, perhaps, blind-from-birth children whose sight has been restored react in an identical fashion, failing to spot that this is a reflection of themselves, although before long, they soon learn to appreciate what the image is actually showing.

What has proved particularly significant is that, by using this independent method of testing, it is the same groups of animals that are believed to show signs of grief (see chapter 3) which also display signs of self-awareness. The list includes all the great apes, including chimpanzees (*Pan troglodytes*), and their close relative, the bonobo (*P. paniscus*), as well as orang-utans and gorillas. No other primates have passed the test, although rhesus macaques (*Macaca mulatta*) might be capable of some self-recognition.

Elephants, too, display self-awareness with the mirror test, and this trait has also been recognised in various cetaceans, where it has been possible to test for such behaviour, in the case of both orcas (killer

Orang–utans can recognise their reflection, as Charles Darwin first confirmed.

Orcas have been shown to display signs of self–awareness.

whales) and bottlenose dolphins kept in marine parks. As far as birds are concerned, magpies have also shown self-awareness when tested in this way, and it has been possible to train pigeons to recognise their reflection, although, unlike magpies, they appear to be incapable of doing so spontaneously. Birds tend to act aggressively at first, though, when confronted by their reflection, which suggests they perceive the image as indicating a territorial aggressor.

Once a magpie had come to appreciate the significance of the image, though, he would often scratch at the image in the mirror, attempting to remove the mark from his plumage. This was taken as evidence that the birds recognised themselves. There is a key difference, however, between birds and mammals in

Dolphins can also recognise each other.

this case, because it is believed that mammals rely on the part of their brain known as the neocortex for self-recognition. Avian species, however, lack a neocortex, and so must rely on a different part of their brain for this ability.

Animal Grief

Although mirrors, as such, do not exist in the wild, animals may encounter various reflective surfaces, particularly water, which can allow them to recognise their own image. However, not everyone accepts the value of mirror-testing, particularly in the case of species which rely less on their sense of vision than do primates. As a result, modifications have been made, working, for example, with urine to test the possibility of self-awareness in domestic dogs.

As far as people are concerned, a specific medical condition known as prosopagnosia — popularly called 'face blindness,' impairs the ability of the brain to recognise faces. Although often the result of brain damage in the vicinity of the area known as the fusiform gyrus, within the temporal lobe, it's also believed there is a congenital form of this disorder. Other characteristics, such as a person's voice, are used to help sufferers learn to identify others.

This means that affected individuals may, in fact, be unable to recognise themselves in the mirror test, but at the same time, there is no doubting their self-

Canine recognition is more scent-based than visual, which makes the standard mirror test less reliable in the case of dogs.

A red squirrel gathering nuts for the winter suggests foresight, but this is misleading.

awareness, which is simply expressed in other ways. The significance of this condition — if any — in the context of animals is still unclear.

Critics of the mirror test also believe that it provides too simplistic an insight into the possibility of animal consciousness. They suggest it confirms that animals which display this faculty simply recognise themselves within the context of their environment, but are not necessarily fully self-aware in terms of consciousness.

This ties in with metacognition, which can be described in functional terms as having the capability to 'think about thought' — and thus be aware of one's existence. The ultimate proof of this is awareness of our own mortality. Studies based on other species have revealed that great apes again are capable of displaying metacognition, as once again may rhesus macaques, but investigations with avian species have not yielded any definitive results to date.

The field of animal consciousness is obviously a very complicated one, with so many variables which make it very easy to misinterpret data. Some actions might be construed as consciousness, but this is not necessarily the case. Consider the situation with squirrels which collect nuts to sustain themselves over the winter. It is easy to suggest that they have a clear insight into the future, but, in reality, it appears that they do this thanks to their instincts, rather than a

Animal Grief

deeper awareness of their being. They carry on regardless, irrespective of the circumstances, whereas if they were aware, they would adjust their behaviour, as people do, by adapting to a particular situation.

This again confirms some of the difficulties confronting an investigation of animal grief, where it is not just behavioural observations that are important, but also correct interpretation of these findings. Luckily, though, as documented later, key anatomical features have recently been identified within the brain, which have given further insight into unravelling which species can feel grief. Interestingly, these tend to be the same as those which display self-awareness in mirror tests, thus providing independent physiological and anatomical evidence to confirm their ability to experience this emotion.

CATHOLICISM RECOGNISES AN ANIMAL SOUL

Over recent years, an increasingly open-minded approach to the belief that some animals do indeed have the ability to grieve when their fellows die has evolved. The theological barriers which have existed over centuries, blocking any serious rational investigation into this field, have broken down.

A major change of emphasis reflecting this shift in religious teaching occurred in 1990, when Pope John Paul II affirmed that

An angry gorilla. As with humans, animal emotions can be indicated by facial expressions.

animals possess a soul, and, as a result, were not to be considered inferior to people. This marked a very major departure from hundreds of years of previous Christian teaching, and most significantly, it gave tacit support to the view that animals could experience emotions, too.

BODY CHEMISTRY

Emotions, generally, are often portrayed as simply an irrational response to a set of circumstances, but, in reality, this is far from the truth. Emotions can bring survival benefits, as typified, of course, by the 'fight or flight' hormone called adrenaline. This is produced in the adrenal glands, which lie close to the kidneys, and like other hormones, adrenaline acts on its target organs through the body, carried there in the bloodstream. Produced in response to a perceived threat, adrenaline helps the body to react in a potentially dangerous situation, by increasing both heart and respiratory rates, as well as serving to provide more energy, by stimulating metabolism.

But it is the hormone cortisol, produced from another part of the adrenal glands, that is most likely to be released into the bloodstream in higher quantities during periods of grief. Cortisol is secreted in increased volume during stressful periods, with raised cortisol levels in people being common in cases of depression.

There is what is known as a circadian rhythm to cortisol levels under normal circumstances, with peak daily output recorded at around 8am and then 4pm, the figure falling back significantly at night. In cases of depression, though, the circulating cortisol level remains consistently high.

Investigations in this area could be useful in providing further insight into whether there is a measurable chemical response to grief in animals. Unfortunately, as yet, however, no detailed studies of this type have been undertaken.

It is not just hormones that can affect an animal's mood and cause what would be described as

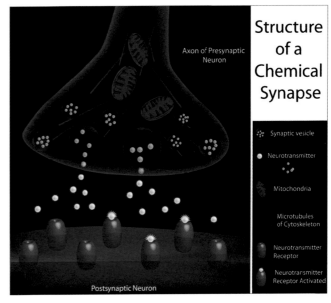

Neurotransmitters play a critical role in the transmission of nerve impulses.

Structure of a Chemical Synapse

Axon of Presynaptic Neuron

Synaptic vesicle

Neurotransmitter

Mitochondria

Microtubules of Cytoskeleton

Neurotransmitter Receptor

Neurotransmitter Receptor Activated

Postsynaptic Neuron

an emotional reaction. Other chemicals in the body, known as neurotransmitters, are known to have an impact on behaviour, too. Neurotransmitters serve to allow nerve impulses to cross the gap, which is known as a synapse, between the neuron or nerve cell and its target cell. In fact, adrenaline itself acts as both a hormone and a neurotransmitter.

Some 30 different neurotransmitters have now been identified, and it is these which act primarily in the limbic system and hypothalamic area of the brain, where emotions are centred, that are most likely to be affected by grief. Neurotransmitters are generally produced and act much more locally than hormones. Dopamine, serotonin, and noradrenaline (also known as norepinephrine) are the most significant in this context.

Serotonin is a very important mood regulator,

Animal Grief

and some human sufferers of depression appear to have lower than normal levels of the chemical compounds resulting from the breakdown of serotonin in their brain and cerebro-spinal fluid. It has been shown that low levels of noradrenaline can be linked with depression, too, and the condition can improve if an antidepressant that boosts levels of this neurotransmitter is taken. But the precise relationships in this area are still unclear in people, quite apart from animals, although such drugs have been used in parrots with some success, as a means of treating depression which can be linked to grief (see chapter 4).

REACTIONS TO GRIEF

Definitions of grief do vary, and frequently, the word itself is often used interchangeably with bereavement, but there is actually an important difference between the two. Grief can be defined as the reaction to death, whereas bereavement is used to describe the actual state of loss.

Grief, as a response, is a complex emotion, operating on different levels. This means that its manifestations are frequently unclear, particularly in the case of animals, where it is essentially impossible to determine their actual feelings, compared with those of a fellow human.

As far as companion animals are concerned, the impact of permanent separation from an owner can effectively have similar symptoms to actual grief. Even though that person may still be alive, the impact can be the same as if he or she had died.

There is no doubt that certain animals, such as elephants, do have powerful memories. There are various cases of individual elephants recognising people they have been trained by and worked with in the past, even if they have not seen them for many years. It is tempting to suggest that the elephant might have suffered some form of grief when they parted company, just as it then expresses signs of excitement at being reunited with that person.

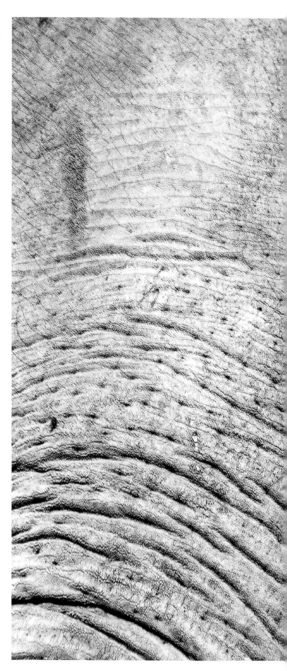

Elephants have long memories.

Animal Grief

There are rituals surrounding deaths in animals, even if these are not as complex or frequently seen as with people. After death, there may be clear recognition of the body of the deceased in some cases, as manifested by an emotion that can only be described as grief, although reactions amongst animals generally are certainly not widespread or consistent.

The emotional reaction may be instant, as in the case of great apes, and yet can also persist, particularly in the case of elephants, who often express a group reverence, when confronted with the remains of a deceased individual — especially one that was known to members of that particular herd.

Grief, typically, has a social dimension, involving others who knew the deceased, in the case of both animals and people. There may be visual indicators of grief, too, with crying being common in people, although this reaction is not normally associated with animals.

Physical changes in the body may be linked with grief. The immune system is often affected, so that we are more vulnerable to colds and other infections following the death of a close friend or family member. This may be linked with associated physiological disturbances such as loss of sleep and appetite, although these indicators are also uncommon in animals.

Overall, however, the most evident aspect of grieving, as far as animals are concerned, is a change in their reactions. This can be seen particularly in the case of companion animals, since they can be observed closely in domestic surroundings, and their regular behaviour is usually well-known, making deviation from it more obvious.

EXPRESSIONS OF GRIEF DIFFER

In Western culture generally, religious rituals help to mark the loss of a person, and they are subsequently remembered in various ways by their community. This is widely considered to be a clear manifestation associated with grief, and since animals do not engage in protracted rituals of this type, this has been commonly used as justification to suggest that they are incapable of grieving.

In this respect, it is very easy to overlook the fact that our own death rituals vary widely, though, perhaps as graphically illustrated by the differing rites associated with various North American tribes. In the Arctic region, it was normal practice simply to abandon bodies on the frozen wasteland, where they would be scavenged and eaten by animals such as wolves and polar bears. This perceived lack of reverence would be regarded as appalling and animalistic in other societies. In marked contrast, the Hopewell tribes constructed lavish tombs for their dead in what is now the upper Midwest region of the USA.

A long period of mourning is also far from universal in human society. In the south-western part of North America, it was traditional for Navajo tribes to bury their dead with very little ceremony, while

Impressive Hopewell burial mounds are still evident today in the landscape of the US state of Ohio.

Apache peoples not only interred corpses very quickly after death, but also burnt the homes and possessions of the deceased.

It's therefore important not to dismiss the concept of grief in animals because their responses do not necessarily correspond to ours, particularly when there is no uniformity within our own species, even. Equally, the ritualisation surrounding the process of grieving is quite distinct from its physiological impact, and many of our most obvious actions when grieving are linked with established social ritual. Grief itself is a very personal — and often private — emotion.

In Arctic regions, polar bears would scavenge human bodies left out on the ice.

Why get emotional?

Grief is a complex emotion, which manifests itself in a number of ways. One of the problems of recognising grief in animals is that we probably tend to interpret it too narrowly, expecting animals to conform to our definition of grieving behaviour. Although the most obvious and instant reaction to the death of a loved one is a conspicuous outpouring of emotion, frequently, there are other, less evident responses, which can also be interpreted as manifestations of grief, and apply in the case of animals as much as with people.

Thanks to the social bonds which bind groups of animals, so a subsequent period of adjustment is needed in their case. The death of the matriarch of an elephant herd will undoubtedly affect the established structure, effectively passing responsibility of leadership to another member of the group. It may be that the emotion arising from the grieving process unites the group, reducing the risk of conflict at this stage, and thereby ensuring a positive outcome for the herd as a whole, helping to facilitate a smooth transition in leadership.

DIFFERENT JUDGEMENTS
Where two dogs have lived together over the course of years, with the younger individual having been

Grief may reinforce the social structure in an elephant herd.

deferential to his older companion, even in this case there are likely to be recognisable behavioural changes when the older dog dies. For example, the younger dog may appear more nervous at first, without his companion to follow, as the established order has broken down.

It's quite possible, and even likely, that many animals — including those where grieving has not been formally recognised in human terms — do, in fact, display grief, based on an acceptance that this is a multi-faceted emotion, covering a range of different behaviours. We tend to view it in human terms, as far as animals are concerned, and the fact they do not display significant, recognisable outpourings of emotion suggests to us that they are generally unaffected by this emotion.

In fact, while grief has understandably come to be defined in many people's minds by its emotional component, a broader understanding would entail a more dispassionate summary of its diverse effects, both on surviving individuals, and — where appropriate — the group as a whole. Grief could simply be defined on this basis of being a change in the behaviour of one or more individuals, following the death of a companion.

Animal Grief

Dogs sharing a household build a relationship.

possibilities of an afterlife, and concentrating instead on the impact on those still alive, it becomes clear that signs of grief are actually much more commonplace in the animal world than previously thought.

The fact that the younger dog does not break down in tears following the death of his companion is not a reliable indicator of whether or not he feels a sense of grief. A clear indication of loss — mirroring the lasting impact on his life that his companion's passing has had — is undoubtedly evident from the change in his behaviour.

A DEGREE OF CONTROL

As mentioned previously, emotions may be seen as random reactions of differing intensity to events outside an individual's control, when, in reality, they are actually measured (and measurable) responses which aid the survival, not just of the individual concerned, but the species as a whole.

When a pack of Basset Hounds chases after a group of rabbits, the rabbits react by darting down their warrens, out of reach. They do not remain at the surface, trying to deter their pursuers by wrestling with them as, of course, it would be impossible for the rabbits to overpower the hounds. Adrenaline — often dubbed the 'fight or flight' hormone, for obvious reasons — triggers this reaction, combined with the rabbits' instincts. Survival dictates that the rabbits must avoid conflict, and use their speed and local knowledge of the terrain to escape their pursuers.

Should one of their number be caught, the other members of the group will not rally round to defend him, and try to help him escape. Instead, the rabbit is left to his fate, which allows the majority to survive. Keeping their emotions in check and not reacting aggressively in these circumstances aids survival of the majority, and the species.

It is impossible to argue that rabbits feel grief

By stripping away the primary emotional response which currently surrounds it, and the associated philosophical debates about the soul and the

Rabbits will seek to escape rather than fight.

aggressively; it simply depends on the circumstances. During the mating period, bucks (males) can be observed fighting quite viciously on occasion, to win the right to mate with particular does. But survival instincts always override grief. In fact, if a doe gives birth to a single offspring, rather than a larger litter, the likelihood is that she may instinctively kill her youngster, rather than waste precious time and resources rearing a single individual, and will soon mate again. On this occasion, hopefully, she will give birth to a much larger litter, fulfilling her reproductive potential, and thereby helping to ensure survival of the colony. In the case of wild rabbits, life is generally short and dangerous, with little room for sentiment. Grieving would serve no purpose.

Another, even more shocking example of this type is well-documented in prides of lions. When a new male assumes control of the pride, it is quite common for him to kill any young cubs in the group being suckled by lionesses. At one level, this seems very strange behaviour, a display of random and extreme violence, given the fact that lions are what can be defined as 'apex predators,' at the top of the food chain, and facing few dangers themselves.

But male lions especially face constant challenges to their status from younger individuals, with the threat in this case coming from their own species. Packs are not stable communities, as the leader is likely to be displaced at any stage, compared to the situation with elephants where the matriarch can lead the herd for many years.

By killing the existing cubs after assuming control of the pride, the male lion will have the greatest opportunity to father his own offspring, before he, too, is displaced in due course. The female may only make a cursory attempt to protect her young under these circumstances, rather than fighting viciously, as they could do, to defend them.

Interestingly, the breeding method of wild cats actually corresponds to that of rabbits – both are

when one of their number is caught by a predator. They will react in the same way, whether or not one of their number is killed, when running for cover. They only emerge again when it is safe to do so. This is not to say that experience does not play a part in survival, because it self-evidently does, in virtually all species.

Furthermore, rabbits are capable of reacting

Animal Grief

Being prepared to fight is a key element of the behaviour of mature male lions, but it leads to uncertainty in terms of pride cohesion, and can result in cub deaths.

defined as being induced ovulators. This means that, instead of undergoing a regular oestrus cycle, as is common amongst mammals, it is the act of mating that serves to stimulate the release of eggs from the ovary, thereby significantly increasing the likelihood of conception.

BIOLOGICAL PARAMETERS AND GRIEF

Females in general give birth to more offspring, or lay more eggs than is required, simply to maintain population numbers. This is because young are especially vulnerable to predators, often because of their size, as well as the fact that they are relatively unaware of the dangers they face. Under very favourable circumstances, if more offspring survive, this enables the population to grow.

Very strong maternal bonds do exist, particularly in the case of those animals traditionally described by zoologists as 'K-selected' species. These are animals which have a long potential lifespan, and will produce relatively few offspring, while displaying a high degree of parental care toward them. Aside from ourselves, great apes, elephants, and cetaceans fall into this category. It's probably no coincidence that grief also tends to be most apparent in members of this group.

This, then, raises the question of what could be the biological advantage — even imperative — of displaying grief?

The ties between mother and offspring in K-selected species group are incredibly strong as a general rule, representing some of the closest relationships in the natural world. Grief helps to create a common bond between group members, and, almost by default, reinforces the need to keep the young safe.

Animals best documented as able to grieve are those which form long–lasting social bonds, such as chimpanzees which can live for 50 years.

Animal Grief

Rabbits, in contrast, are regarded as r-selected species, where rapid reproduction is the key to survival, with individuals living relatively short lives.

PARENTAL BONDING, AGGRESSION, AND EMOTION

Bonding between mother and offspring derives in part from hormonal ties in the early stages after birth, and is linked with lactation in the case of mammals. This can impel a mother to endanger herself in order to protect or save her offspring, in a way that would otherwise not be expected; even taking on a formidable predator. Trying to come between a mother and her young can prove dangerous, even if, under normal circumstances, the female is quite docile.

There are numerous examples of this type of behaviour which can be seen in the form of videos posted on the internet, many of which have been filmed on the plains of Africa, where the battle for survival between the larger grazing herbivores and associated predators such as lions is especially intense.

Communication between mother and young can also be remarkable, even in the mayhem of mass migration. A female wildebeest (*Connochaetes* species) can pick out her youngster if they become separated, although even if it succumbs to an attack by predators, she will display no obvious signs of grief, continuing on her hazardous journey alone.

Nor does this apply just to mammals. Various species of birds are well documented as acting not just aggressively, but even combining, in mixed species groups, to protect their offspring. In a woodland setting, nest-robbers such as jays may be driven off in this way, and formidable predators such as snakes harried in other cases.

In North America, the curve-billed thrasher (*Toxostoma curvirostre*) is well-known for launching attacks on California kingsnakes in order to protect

Young elephant calling.

A zebra lashes out at a lioness with her powerful hind legs, in an effort to keep younger members of the herd safe from attack.

A protective nature is common in many animals, even if they are not recognised as having a sense of grief. Here, a group of buffalo form a distinctive ring around a calf.

Animal Grief

Mother and calf cross Kenya's Mara River safely together, in sight of a watchful crocodile, on the annual wildebeest migration.

her brood. Once again, though, these are displays of instinctive, protective aggression. If the bird's attempts fail, she will show very little, if any, appreciation of loss, aside from perhaps remaining briefly in the vicinity of the empty nest afterward. Indeed, she may well start nesting again in a short space of time; any prolonged period of grieving would be harmful, by deterring reproductive activity.

GRIEVING STUDIES
Studies of grief in people suggest there are several distinct steps in this process, according to the research of Swiss-born psychiatrist Elisabeth Kübler-Ross (1926-2004). In her book, *On Death and Dying*, published in 1969, she set out five recognisable stages.

There is an initial sense of shock, which can border on denial, even if the death is not unexpected. This is thought to insulate against being emotionally overwhelmed. Reactions of this type have been noted in animals, too, particularly in the case of mothers who have lost their young. It can be manifested in various ways, with elephants often trying to raise their dead,

The curve–billed thrasher will defend her young from California kingsnakes.

while great apes will carry around their dead young for a time. Even animals not noted for grieving, such as lionesses, may pick up a deceased cub, if it has died unexpectedly.

A similar set of feelings can affect owners who have lost their companion animals suddenly. The initial sense of shock may well be followed by a period of pain and guilt. This is often completely illogical, on the grounds that nothing more could have been done for the animal, but can be overwhelming for a time, even affecting the ability to carry on with normal daily life, particularly when confronted by reminders of a beloved animal that is no longer there.

Such feelings may be linked with a sense of anger, too; seeking someone to blame for what has happened, even if you know deep down that this is not reasonable. Nevertheless, you may convince yourself at this stage that perhaps the veterinarian should have been able to do more, for example, or that if the driver

had been paying attention, your cat would not have been run over, even though you know that she actually ran straight out in front of the vehicle, and there was no chance to brake and avoid hitting her.

A sense of melancholy and loneliness then descends, and you may find that you want to be reminded of your friend, by looking at old photographs, perhaps. This is the stage at which the true magnitude of your loss becomes apparent. Unfortunately, it is very easy to become trapped in past memories at this point, rather than looking to the future.

It's important to realise that grief can become pathological – acting as a cause of ongoing depression, particularly if you have sustained other losses recently in your life – thereby engendering a sense of hopelessness. You may turn away from friends, and carry on, unable to put the past behind you.

It can be even harder when it comes to grieving for an animal rather than a person, simply because

Gorilla mother and baby.

of the commonly-held view that grief for an animal cannot be as intense, although in reality, this is very often simply not true. This is not something that can be easily predicted, but depends on a variety of factors.

The fear that others will react by saying you will 'soon feel better,' and that you should get a new companion, will not help resolve the underlying trauma you will be experiencing at this time. Thankfully, however, there are now bereavement counsellors trained to help people who are experiencing difficulties in terms of overcoming the loss of a companion animal.

There is no set time frame when it comes to grieving, as it is very much a personal response, and depends on individual circumstances. In due course, though, you should find that, as life continues and you develop a different routine without your companion, you will start to feel less sad and bereft. Hopefully, before too long, you will be able to remember the

Young elephants are watched over by other adults in the herd, not just their mother.

happy times that you enjoyed together, and take comfort from these.

DO ANIMALS GO THROUGH STAGES OF GRIEF?

It is obviously impossible to determine whether animals feel grief in the same way if a close family member dies, or they are split up. There is indisputable evidence to show that dogs especially can react in a negative way to the loss of a companion, or a change in residence though, so it is quite likely that they do.

Grief can operate at various levels. Of course, it's possible to grieve not just over a death but also a separation from a loved one, a move from familiar

33

Animal Grief

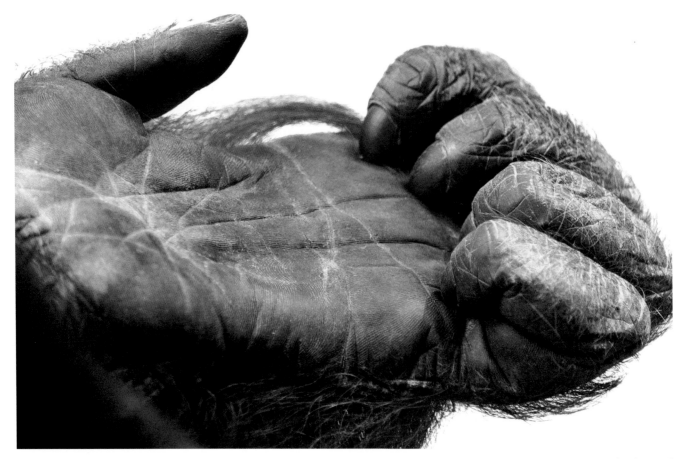

Physical contact amongst great apes, whose hands have fingerprints like ours, has been shown to provide reassurance in times of stress and grief.

surroundings, or, in fact, anything that may engender a strongly negative and personal emotional response.

Generally, the degree of grief that animals suffer does not seem to be as intense as that which people experience, nor do they deal with it in the same way, in terms of breaking down and revealing their feelings.

Alternatively, the intense grief experienced by people could be a direct reflection of the emphasis which we as a species place on verbal communication. In order to overcome our feelings of grief, it can be very helpful to talk about them. Animals, by way of contrast, rely to a much greater extent on a range of senses, with verbal communication not being as significant.

A mother chimpanzee giving affection and reassurance to her youngster.

NEW THINKING

More recent research by clinical psychologist Professor George A Bonanno, based at Teachers College, Columbia University in New York, has actually raised serious doubts about Kübler-Ross' step-by-step progression model of grief, based on the fact that research has confirmed that a surprising number of people do not grieve in this formalised way, if at all.

Contrary to what had previously been believed, this so-called resilience, or 'absent grief' is no more pathological than common grief, where sufferers do display overt and recognisable signs of grieving. Most

Animal Grief

Elderly chimpanzees: these two are likely to have spent most of their lives together, reinforcing the bond between them, and the sense of separation and grief when one of them dies.

remarkably of all, though, Bonanno established that 'absent grief' was the most widespread form, recorded in half of all people, alongside common grieving, and long-term chronic grieving.

This is potentially a highly significant finding as far as understanding animal grief is concerned. Researchers in this area have tended to concentrate on trying to establish distinctive, recognisable patterns of behaviour in individual species as a means of identifying animals which may experience and show grief.

In reality, though, based on Bonnanno's research, it's clear that there is no uniformity in reactions to death amongst people, let alone in animals, as evidenced by individual reactions. While a chimpanzee female may appear distraught by the loss of her youngster, repeatedly trying to revive it, other members of the group are unlikely to respond in such an emotional way to the baby's death.

What they have been observed as doing under these circumstances, however, is to try and comfort the bereaved mother, grooming and touching her more frequently, almost as a way of providing reassurance. Interestingly, the sense of loss tends to be felt more intensively through the group when an adult member dies, and perhaps this is a reflection of individual contacts and relationships through the years, rather than the very brief exposure which the group will have had to the newborn infant.

IS IS POSSIBLE TO DETECT GRIEF?

A new scientific method has recently been used to investigate the grieving process in terms of changes in the brain. Functional magnetic resonance imaging (fMRI) relies on detecting shifts in blood flow, in response to neural activity. When energy is being used in a particular area of the brain, so blood flow increases very rapidly — within five seconds — to that particular part of the brain. This is because the neurons require both oxygen and glucose in order to function.

This method has provided a way to recognise the physiological mechanisms which indicate that grief may be felt more acutely by some than others, and the areas of the brain involved. Furthermore, such studies have provided important insights into the way that these reactions are controlled at a cellular level in the brain, and the inter-relationships between the different parts of the brain.

The area known as the amygdala is believed to be linked to memories built around emotional events, a fact confirmed in the fMRI studies, which revealed that people who reported feeling rather overwhelmed by thoughts of grief showed increased activity in the ventral amygdala, as well as the anterior cingulate cortex, another part of the brain involved with emotions and empathy, as well as associated physiological impacts on body processes, such as heart rate.

Signs of aggression can be seen in greater apes, like this older chimpanzee. Such behaviour is usually accompanied by menacing vocalisations. Aggression is a much easier, and more straightforward emotion to recognise than grief.

Although, at present, no studies have been carried out using fMRI to investigate possible brain manifestations of grief in animals (and they could be hard to accomplish), this area of investigation may in the future offer distinct pointers to emotional reactions in various species, providing clear indications of their feelings.

Wild observations

Stories of wild animals displaying emotions are not new or unusual. When the whaling industry was at its peak, there were a number of accounts in ships' logs of how whales would ram vessels when one of their number was killed, oblivious to their own safety. One particularly well-documented and interesting incident of this type took place on November 20, 1820, in the South Pacific.

The *Essex*, a 238 ton whaler, was pursuing a pod of sperm whales (*Physeter macrocephalus*). These leviathans of the deep can grow up to about 20m (60ft) long, and possess the largest brain of any animal alive today, weighing up to 9kg (20lb).

Most of the *Essex's* crew were in the ship's three smaller long boats when the drama began, having just struck one of the whales in the group with a harpoon. The injured cetacean had holed one of these boats with its powerful tail flukes as it thrashed around, forcing the sailors onboard to row back to the *Essex* for repairs.

EXPRESSIONS OF ANGER AND GRIEF?

Just after the damaged boat had been hauled aboard

Abandoned long boats left at a long-forgotten whaling station in Antarctica.

the *Essex*, the first mate, Owen Chase, then noticed that a large sperm whale had separated from the pod and was swimming at full speed towards the ship. The collision was massive, occurring just behind the bow, and it caused the *Essex* to start taking on water immediately. The crew onboard signalled urgently for their companions to return, and started working desperately in an attempt to prevent the vessel from sinking.

The whale, meanwhile, remained at the surface some distance away for a time, twisting and convulsing, before regaining its focus and heading back determinedly towards the ship once again. This time, its head was clearly visible above the water as it crashed directly into the port side, and there was to be no escape. The *Essex* began to submerge rapidly, forcing those aboard to take to the already damaged long boat, after rescuing some supplies, before the ship finally sank.

The whale that caused the damage was not seen again, and was presumed to have been fatally injured. What may have triggered its aggression was not just the attack on the other whale, but the fact that two other members of its pod had been killed by the crew just beforehand. Although it was not known at the time, sperm whales can communicate with each other by means of whale song over considerable distances through the water.

The whale probably did not need to see that its companions were being killed – doubtless it was well aware of what was happening. Possessing a natural lifespan similar to if not greater than our own, these cetaceans would also remember past encounters with whaling ships, helping explain why they might react aggressively under such circumstances.

There is no doubting the bravery of sperm whales, either. They are known to engage in titanic battles with giant squid in the depths of the oceans, their bodies often bearing the scars of such encounters, inflicted by the squid's numerous tentacles. The way that the individual involved in the encounter with the *Essex* reacted was therefore not entirely surprising, displaying, as he did, both anger and grief, two emotions that can be closely allied, and not easily separated.

As for the *Essex's* crew, it was left in a desperate situation, although, luckily, Owen Chase's boat was picked up by another ship, while the captain made it to the safety of land, reaching Santa Maria Island, off the coast of Chile. Those in the third boat were not so lucky, however. After a storm, the group became separated and they simply disappeared.

Owen Chase wrote an account of this experience, and, undeterred by his father's experience, his son,

The bond between a female whale and her calf is very strong. Here, a humpback whale (Megapytera novaeangliae) is shown with her calf.

The sight of a whale breaching — the term used to describe jumping out of the ocean in this way — is an awe–inspiring indicator of power. Life was hazardous for early whalers, as well as the whales themselves.

Animal Grief

William, also became a whaler. An interesting postscript to this event is the fact that William Chase described these events at length to a fellow whaler, and passed him a copy of his father's book. The man's name was Herman Melville, who would ultimately write the best-known account of whaling, involving a sperm whale, in his novel *Moby Dick*.

OTHER ENCOUNTERS

While the case of the *Essex* is probably the most famous story of its type, there are various other accounts of whales using their considerable strength to attack whaling vessels. Collisions between the smaller boats sent out to harpoon the whales and their quarry were not uncommon, because of the nature of the conflict, but some whales learnt that attack could represent the best form of defence for them and their companions.

One of the most formidable and legendary members of this group was nicknamed Mocha Dick by whalers. This particular individual was named after Chile's Mocha Islands, where he was first sighted. Mocha Dick had an unmistakable appearance, being of a distinctive pale grey colouration, and with a long white scar, measuring about 2.5m (8.25ft), on his head.

Mocha Dick is credited with sinking dozens of small whaling boats, as well as attacking many larger whaling ships, possibly accounting for over 100 vessels in total over a period of nearly half a century, from 1810 until 1859. One account, however, suggests that he was killed in 1838, defending a cow whose calf had just been taken by a whaler.

MISPLACED AFFECTION

Even today, although large-scale whaling is outlawed by international consent, cetaceans are facing new threats in the oceans. It is possible, for example, that sonar – the underwater navigational system used by submarines and other vessels – may disrupt their means of communication, and could disorientate them, leading them into danger.

Whales do not display the anger and ferocity towards whale-watching cruises that they sometimes used to display towards whaling ships, having come to recognize the danger that they represented.

Around the world, strandings of cetaceans appear to be on the increase, with individuals becoming trapped in shallow water. There are various other possible reasons for such behaviour, of course, and some events of this type can be triggered by illness. The loyalty of a pod of whales may prove to be so strong that if a sick individual heads to shallow waters, the other members of the group will not abandon their companion, and will follow him or her, thereby ending up becoming beached as a result.

This behaviour is especially common in pilot whales (*Globicephala* species), but has also been recorded in a variety of other species, including sperm whales and dolphins. It is obviously impossible to state categorically that the whales behave in this way out of grief for their ailing companion, but there is undeniably an emotional response behind it.

In fact, when faced with a mass stranding of a pod of pilot whales, researchers have found that concentrating on taking the young members of the group out to sea is the best way to encourage the others back to deeper water. This, again, is the result of an emotional response, because hearing the distress calls of the youngsters evokes a very strong, protective reaction amongst the adult members of the pod. These particular calls are identifiable variations of their normal whistles, used to indicate that the whale is in danger.

This very close bond with the young members of the group may stem from the fact that pilot whales breed more slowly than most other whales, with females perhaps calving only once every five years. Emotion, therefore, helps bind the group together, ensuring that the effort that goes into protecting the

The remains of a whale carcass on the Antarctic coast, surrounded by penguins. Sick and dying whales may venture into shallow water.

offspring is maintained. Under certain circumstances, however, this also, sadly, has the potential to lead the pod into danger.

MEMORIES AND GRIEF

Animals that tend to display signs of grief are those which are known to have memories. There is a popular saying – 'an elephant never forgets' – and, actually, there is a reason for this, as is apparent from a study of the elephant's brain. The temporal lobes here are highly developed, and their folded appearance means that

these pachyderms possess a remarkably good memory. Furthermore, as with cetaceans, elephants have their own long-distance system of communication, being able to engage with other elephants up to 20km (12ml) away, relying on the use of infrasound for this purpose, which is inaudible to our ears.

There is also no doubt that elephants recognise each other, both in life and after death. Their normal contact sounds are as individual as are our voices, and scientists have found that elephants will react to recordings of the sounds made by dead companions,

In some cases, mass strandings of ceteaceans may be the result of misplaced loyalty within a group.

confirming that they have a memory which relates to individuals.

Elephant herds comprise females with their youngsters and juveniles. Members of the herd are very protective towards the younger members of the group, which is led by an old individual, known as the matriarch. In times of drought especially, the herd relies on the matriarch's memory, which contains detailed knowledge of their territory, in order to help them find water as well as food.

The whole group will also be alert to potential dangers that the younger members may not appreciate. Their size means that young elephants are more vulnerable to predators than adults, with lurking crocodiles in the shallows when they are drinking representing a particular danger.

Various documented cases describe how adult elephants will react very violently to save a calf attacked by a crocodile. They will use their weight to stamp on the reptile, and on one occasion, when a crocodile was ill-advised enough to react by seizing an adult elephant by its trunk, the reptile ended up being slung high into a tree for its trouble, sustaining fatal injuries as a result.

ELEPHANT GRAVEYARDS

It used to be widely believed that elephants would leave the herd when they knew they were going to die, and this gave rise to stories about elephant graveyards. Studies have revealed that there is some truth in these accounts, because, as with various other herbivores, so the teeth of elephant simply wear down as they grow older.

Elephants grow six sets of teeth in succession during their lifespan, which can be of equivalent duration to our own. Ultimately, as the final set is worn down, it becomes progressively harder to eat, leaving the animals facing starvation. As a result, they are drawn to lush vegetation, which is softer and requires less chewing, and plants of this type are most common around waterways and in swampy areas. Consequently, it is here that older elephants are ultimately likely to die. Explorers and ivory traders often encountered their remains in such areas, which fuelled the belief in stories about elephant graveyards.

Equally, though, elephants do display an evident understanding of death and grief, both in the case of their own species and others, too. There are a number of accounts of how a herd will stay with the remains of a companion, particularly if it is the matriarch, gently touching her body with their trunks. Given the sensitivity of the trunk, this may be a means of

Animal Grief

Elephant bones are of special significance to herds, and are treated with reverence, suggesting some sort of grieving ritual. Note the tracks of other elephants around the bones.

Elephant herds have a well-defined social structure, with individuals recognizing each other.

Animal Grief

recognising the demise of the individual in some way, or alternatively, an offer of help, trying to rouse the fallen elephant.

When this fails, the elephants often begin to throw leaves and other vegetation in the area over the body, along with soil, almost as a token of respect. They will then continue to spend some time in the vicinity of their dead companion, leaving only for food or water as necessary, before returning – sometimes over the course of several days.

There is also a remarkable account by George Adamson, whose work in Kenya with his wife, Joy, gave rise to the book and film *Born Free,* about how elephants can recognise the remains, following the death of one of their group. On one occasion, he was called to deal with a herd of elephants that was breaking into a compound, and was forced to shoot one.

He gave some of the meat to local people, and moved the remains of the carcass over 800m (0.5ml) away from the area where the elephant had been killed. By the following morning, however, the surviving herd members had carefully moved part of the skeleton back to the original site where George had shot the elephant.

The remains of elephants do seem to be of particular significance to others of their kind, and treated with due reverence. The elephants will gently touch the bones with their trunks, and may roll them with their feet as well, even if the remains are of individuals that were unknown to them.

A PHYSIOLOGICAL EXPLANATION

The most significant breakthrough yet in our understanding of whether certain animals can grieve came about in 1995, thanks to the research of Patrick Hof, a neuroanatomist based at the Mount Sinai School of Medicine in the Manhattan district of New York. He was examining an area at the front of the brain, known as the anterior cingulate cortex, and spotted some very distinctive cells there, which had an unusually elongated shape. They were about four times bigger than the neighbouring cells, and had only one connecting branch, or 'dendrite.'

It transpired that the presence of these unusual brain cells had been first documented back in 1881. However, these neurons had been largely overlooked until Hof became interested in them, although a Viennese scientist called Constantin von Economo had described their presence in human brains in 1926, and they have since been named after him. (They are also known as spindle neurons, because of their appearance.)

Apart from being present in the anterior cingulate cortex, von Economo had also found the neurons in the frontal insula. Both these parts of the brain have subsequently been highlighted as being highly significant in the context of our emotions, thanks to recent brain-scanning studies.

Meanwhile, John Allman, a neuroscientist working at the California Institute of Technology at Pasadena, was investigating this field, on the basis of comparative anatomy. He has a particular interest in the development of the human brain, and how this affects social behaviour, with grief itself, of course, being a manifestation of social awareness and links. Thanks to his studies, Allman built up a large collection of brains of different primates, and as Hof's research started to extend into the comparative field, so the pair began to collaborate in their work. By 1999, they were able to confirm that von Economo neurons were present in the brains of all great apes – gorillas, orang-utans, chimpanzees, and bonobos (forest chimpanzees).

Just as significantly, though, the scientists were also able to rule out the presence of these cells in other primates such as lemurs. This, in turn, suggested that the brain changes which had led to the development of von Economo neurons had taken place probably about 13 million years ago, well before our own species diverged from chimpanzees – an event believed to have occurred around six million years ago.

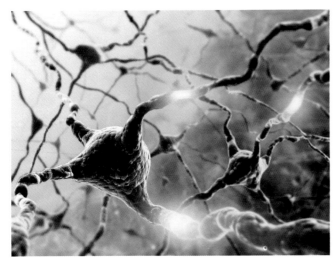

Neurons in the brain are responsible for transmitting nerve impulses.

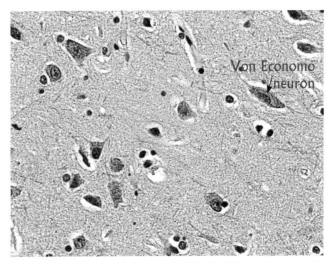

Specialist von Economo neurons, linked with emotions, can be seen here at high magnification in brain tissue.

Rock hyraxes may be social, and are regarded as the closest relatives of elephants, but they lack von Economo neurons, and show no sign of grieving behaviour.

Animal Grief

In spite of what can be a rather humanoid appearance, no von Economo neurons have been identified in lemurs such as Verreaux's sifaka (Propithecus verreauxi).

Von Economo neurons are far more numerous in humans than chimpanzees, which may well be a reflection of the respective difference in brain size, as our brains are typically three times larger.

For a period, it was thought that only humans and great apes possess von Economo neurons, but these cells are now known to be present in the brains of various whales, who developed them at a much earlier stage in history. Their presence helps explain the emotional response of grief to the loss of a companion, and then anger toward those responsible, as witnessed by the crew of the *Essex*.

ONGOING STUDIES

The search has also continued for the presence of such cells in the brains of other animals. Studying the brain of Simba, an elephant who died some years previously at the Cleveland Metroparks Zoo, Allman discovered that these highly distinctive neurons could be identified in the tissue here, too.

Elephants have large brains, and out of a total of some 1.3 million neurons which make up the frontal insula on the right side of the brain, around 10,000 of them – equivalent to 0.8 per cent – are von Economo neurons. This contrasts with an equivalent figure of 1.25 per cent in people.

Comparative studies have revealed that none of the elephant's nearest surviving relatives possesses them, however, and investigations carried out on the brains of over a hundred different species have revealed that they are confined to primates, elephants, and various cetaceans. Professor Allman believes their development corresponds to the increased brain size of these species.

Bigger brains need faster ways of communicating information between the different areas, otherwise the system is at risk of breaking down. In effect, messages have to travel over longer distances, so increased communication speed is critically significant.

Allman believes that the origins of the von Economo neurons can be traced back to a set of neurons located in the insular cortex of the brain, which are present in all mammals, and serve to control appetite. But even if their role was originally simply to control one aspect of behaviour, there is now incontrovertible evidence to suggest that, in specific cases, they have evolved to the extent of actually hard-

Bottlenose dolphins (Tursiops species) are one of the marine mammals now known to possess von Economo neurones.

Large brains may help to explain the development of von Economo neurons in primates and other groups, and thus the ability to feel grief.

Animal Grief

wiring together individuals of the same species in terms of communication and emotion.

The role of von Economo neurons is now believed to have developed into a means of ensuring social recognition, helping to create a bond between individuals living in the same group. There are still many unanswered questions, though, such as how they might be affected by hormonal influences that will trigger a mother to defend her offspring.

It could be argued that more effective communication is important within a group, because members need to work alongside each other, and anticipate what others may do, or how they could react, in order to work effectively in collaboration.

These neurons allow the brain to process visual and auditory information, focusing on group interaction.

There is remarkable footage of a group of elephants which Professor Allman uses to illustrate this point. A young elephant is at risk of drowning in a water hole frequented by the herd, having become trapped by the mud. The matriarch of the group, accompanied by a second female, wade in to save the youngster, pulling and pushing it free using their considerable strength.

A third member of the herd then begins to trample the earth at the edge of the waterhole, to provide an easily accessible route which allows the calf to reach firm ground. The elephants worked together,

Baby elephants have a long adolescence, and can find themselves in danger, to the extent that they must be rescued by older members of the herd.

Younger members of a herd of elephants depend largely on the matriarch's memory for their survival when food and water is in short supply.

almost instinctively knowing the role that each needed to perform in order to save the youngster, when confronted by this dangerous situation.

Remarkably, the best evidence for the role of these neurons comes from human medicine. A further piece in the jigsaw of understanding their role fell into place when Professor Allman was lecturing to an audience which included a neurologist called William Seeley, who was investigating an illness called frontotemporal dementia (a degenerative process which affects the frontal part of the brain). Listening to Allman talk about von Economo neurons, Dr Seeley realised that the areas of the brain that were affected by this condition were the same as those where these nerve cells are to be found.

Neuropathology, therefore, offered a way to

Animal Grief

provide an insight into their role, if they turned out to be the cells affected by this condition. Working together, Allman and Seeley then discovered that in patients suffering from frontotemporal dementia, it was the von Economo neurons that were largely destroyed. As a result, the behavioural changes seen in affected patients could therefore provide a vital insight into what exactly the role of these cells would be under normal circumstances.

A range of cognitive functions are affected in cases where people are suffering from frontotemporal dementia, with both judgement and social interaction being impaired, as well as emotional awareness. This provides the clearest indication yet that von Economo neurons have a higher cognitive function, helping not only to create a sense of loyalty, but also providing the ability to empathise with others forming part of the group to which the individual belongs.

MOURNING THE DEAD

Elephants have phenomenal memories, which undoubtedly aid their survival in the wild. This is reflected by the greatly enlarged area of the brain known as the hippocampus. In times of drought, when not just water but also food is likely to be in short supply, the matriarch of the herd will use her memory to guide the group to areas where water may still be accessible, with food nearby.

When it comes to recognising their dead, some interesting results have emerged, as the result of a series of studies on the reactions of wild elephants to skulls in the Amboseli National Park in Kenya, by a team led by Dr Karen McComb. When confronted by an elephant skull, and skulls from a buffalo and rhinoceros, the elephants concerned were particularly attracted to the skull of the elephant. They smelt and touched it using their trunks, and also gently stood on it.

But in another test, when confronted with the skull of the matriarch of the herd who had recently died, and skulls from other matriarchs, the elephants did not appear to display any particular preference for her remains to those of the others, which would seem to weaken the argument that elephants regularly visit the bones of relations who have died in their territory, although they are likely to remember where these remains are located.

But a particularly remarkable and compelling piece of evidence of what could only be described as grief – in the context of recalling a deceased individual with affection – has been documented by elephant researcher Cynthia Moss. She had brought the jawbone of a dead elephant back to her camp, when coincidentally, other members of that elephant's family visited the area several weeks later.

The group seemed to know instinctively that it was one of their own herd members: indeed, the calf of the elephant in question stayed behind the group, turning the skull over with her trunk and feet, as if lost in memories. Dr Moss speculated that, at seven years old, this young elephant may have been able to recognise her mother from the form of her jaw bones.

CRYING

One of the most interesting observations, at least to us, is the concept of 'elephant tears,' describing the secretions from their eyes that often roll down their cheeks as they approach another elephant's skeletal remains. There is no doubt that some elephants do seem to undergo a specific physiological and emotional response under these circumstances, which is highly suggestive of grief.

Tear fluid normally helps to keep the surface of the eye moist, preventing it from drying out. This fluid usually drains out of the eye via the nasolacrimal, or tear duct, which is an opening just inside the corner of the lower lid, providing a conduit down into the nasal cavity. The production of tear fluid does not normally exceed the volume that drains away through this opening. A physical or chemical irritant can cause

A tear can be seen running from the eye of this young Asian elephant.

resulting from grief or similar emotional trauma, as they have a higher protein content. Unfortunately, however, it appears that there have been no tests carried out on the tear fluid of elephants, to determine whether they, too, produce different types of tear fluid, depending on the circumstances.

In the case of primates, they may make wimpering sounds, rather like those heard during crying, when they are distressed, but the production of tears of grief does not occur. There is one well-documented exception, though, which was observed by the well-known gorilla researcher, the late Dian Fossey.

She recorded the case of Coco, a young mountain gorilla (*Gorilla beringei beringei*) who had been orphaned by hunters and subsequently kept in a small crate for a month before being brought to her. When Dian let Coco out into a room from which the nearby forest was visible through a window, the gorilla started crying, with tears rolling down her face. It was not something that Fossey ever observed again, and it may have been a reaction to suddenly being exposed to the light, although, understandably, it could be supposed that these were tears of emotion.

SOMETHING MISSED?

The reaction to the bones of their ancestors is particularly unique to elephants and humans, although there is a very interesting account about the behaviour of a now-extinct marine species, called Steller's sea cow (*Hydrodamalis gigas*). It was the largest member of the Order Sirenia, a group which also includes manatees (*Trichechus* species), and the dugong (*Dugong dugon*).

Steller's sea cow was a much bigger species, though, potentially growing to 9m (29.5ft) or more when adult. It spent all its time in the ocean, feeding on kelp – a form of seaweed. Although it is originally believed to have had a wider distribution through the northern Pacific region, the species was already in decline when it was first described by Georg Wilhelm Steller in 1741.

the output of tear fluid to increase, however, resulting in fluid overflowing down the cheeks and giving the impression of crying.

Yet there is a key difference in the chemical constituency of human tears, shed in response to a piece of grit entering the eye, for example, and those

Animal Grief

A very rare skeleton of a Steller's sea cow, on display at the Muséum national d'Histoire naturelle, in Paris, France.

Steller was a German naturalist on a voyage of discovery being led by Vitus Bering, a Danish explorer in the service of the Russian navy. The expedition encountered Steller's sea cow in the vicinity of the Commander Islands, which represent the westernmost group within the chain of the Aleutian islands, lying off the western side of Alaska.

Fewer than 1500 of these inoffensive, slow-swimming mammals were estimated to be alive by the time they were discovered by the expedition. Stellar records a very poignant episode in his journal about this family-orientated, monogamous species, in which an adult pair would invariably be found living in the company of an older offspring, and a younger individual.

Steller noted that after the crew of his vessel had killed a female sea cow and her body washed up on a nearby beach, her mate returned regularly to the vicinity of her inanimate body over the course of two days, acting in such a way "as if he were inquiring about her."

What was not known at that stage, however, is the very close relationship that existed between these marine mammals and elephants, which may well explain the similarity in behaviour, with sea cows visiting their dead in the same way as do elephants.

Sadly, however, these aquatic herbivores became extinct around 1768 — barely a quarter of a century after they had first been recorded by western science.

Therefore there is no way of knowing with any certainty if they had a similar brain anatomy to elephants, complete with von Economo neurons, and routinely showed emotional attachment to dead companions.

CLOSE QUARTERS

Chimpanzees and other great apes will display grief

A dead whale on a beach. In common with most animals, cetaceans do not stay with or return to the bodies of group members.

when one of their group dies, but take no interest in their remains, although mothers can become very distraught following the loss of their offspring. A female gorilla called Gana, living at Allwetter Zoo, Münster in Germany, lost her male baby called Claudio when he was just three months of age. The two had been very close, whereas Gana had actually rejected her first baby, Mary Zwo, when she was just six weeks of age.

Just as happens in the wild, Gana persisted in carrying Claudio's body around with her. It is believed that he died from a heart problem, and initially, she seemed to try to revive him, by shaking and prodding his body. When that failed, she put him over her back and started walking around, pausing to check if he was breathing.

This type of behaviour has also been observed amongst wild gorilla families in Africa, although reactions differ. Young babies which die are often carried for days by their mothers, who frequently appear to be depressed, based on their body language, when compared with other family members of the group.

They may also remain with dying individuals, even after their deaths, for a day or so, rather like keeping a vigil, before moving on, although in some cases, no attention is paid to the body of their companion.

Similar behaviour has been documented in chimpanzees, too, with the female often trying in vain to resuscitate her dead infant, and then carrying her offspring around with her for several days, or returning to the dead infant, gently touching him, in the hope of eliciting a reaction.

TALKING ABOUT LOSS

Perhaps the most dramatic and moving account of grieving in a gorilla was documented in the case of Michael, a silverback who had been orphaned when his mother was killed in Cameroon. He was brought to Stanford University in California when he was 3½

Female chimpanzees will often display clear signs of grief if their infant dies. A forest chimpanzee or bonobo is shown here.

Animal Grief

years old, and joined Koko, who was taking part in a scheme to teach gorillas to have conversations in sign language.

Gorilla Foundation researchers were amazed when Michael described the death of his mother at the hands of poachers, using this new method of communication, revealing that her neck was cut, and the distress and grief that he felt. He repeated this story several times during his life, seemingly relieving the trauma. He died in April 2000.

Trainer Francine Patterson has been instrumental in pioneering this communication scheme, and has had great success with Koko, who was born at San Francisco Zoo. This female gorilla can communicate by means of over 1000 signs, using American Sign Language (ASL), and is believed to comprehend the meaning of more than 2000 spoken words. Koko does not use sentences for communication when signing, but simply relies on individual nouns and adjectives.

Gorillas are unusual, because, like us, they will care for companion animals. Koko, who is now 40 years old, asked to be given a kitten in the summer of 1984. She chose a tail-less kitten, who became known as All Ball, and displayed considerable affection towards her feline companion, treating her very gently, almost like a substitute baby gorilla.

Tragedy struck before long, however, when All Ball slipped out of Koko's quarters, and was killed as a result of being hit by a car. Francine Patterson reported that when it was explained to her what had happened, Koko expressed her grief, by signing words including 'sad,' 'bad,' and 'cry.' She was also later heard mimicking the sound of a person crying.

AVIAN CONUNDRUMS

Whereas feelings of grief in mammals can be communicated directly in some cases, and are also now explicable in terms of neuroanatomy, it is much harder to account for long-standing suggestions that birds, too, are capable of displays of grief. It is

Gorillas can learn to communicate by means of sign language, and may use signing to express feelings, not just of grief but also of happiness.

Albatrosses will pair for life, and may live for over half a century.

probably fair to say that, at this stage, beliefs in this area are more representative of a combination of folklore and anthropomorphism, rather than scientific understanding.

Certain groups of birds, notably waterfowl such as swans and many seabirds, including albatrosses, are known to pair for life. This has been confirmed by a combination of studies, including banding (ringing), and physical identification, such as the bill patterning of Bewick swans (*Cygnus columbianus*) for example, which allows individual birds to be recognised.

So strong is the bond between members of a

A pair of Bewick swans prepare to land at Slimbridge. Each has a unique facial patterning, helping scientists to study their behaviour.

Swans, such as these mute swans (Cygnus olor), typically pair for life.

Animal Grief

pair in the case of this species that staff at the Wildfowl and Wetlands Trust in Slimbridge, Gloucestershire, in the south west of England, believed that disaster had befallen a male Bewick swan christened Sarindia, when his partner, Saruni, arrived at the sanctuary for the winter of 2009-2010 with a new mate, after migrating from the species' traditional breeding grounds in Arctic Russia.

Remarkably in this particular case, however, Sarindia arrived later, but there was no interaction between the members of this former pair. Their split marked only the second occasion in the 40-year history of the reserve, which is home to 8000 swans over the winter period, that a change of partner has been observed.

Under normal circumstances, a swan will only seek a new mate if the existing partner has been killed. Swans are long-lived birds, with a potential life expectancy measured in decades, although they face an increasing number of hazards today, particularly when they are migrating.

It is tempting to suggest that because the birds develop such a strong bond in life, so will it be transposed into death, when a member of the pair dies, but this does not appear to be the case. Similarly, swans can be particularly aggressive in defence of their eggs or young, potentially putting themselves in harm's way, as they hiss and lunge, as well as flapping their powerful wings. If the defence fails, however, and their cygnets are killed, the birds do not display any obvious signs of grief.

No records seemingly exist of swans being drawn to others of their kind who have died, despite the fact that these waterfowl may make a very distinctive, readily identifiable sound when they die, giving rise to the term 'swansong' as a way of describing someone at the very end of their career or life.

There are cases where other birds do flock around others of their own kind, oblivious to danger, putting themselves at risk of being killed. This is a technique that has been exploited by hunters, and led to the extinction of the passenger pigeon (*Ectopistes migratorius*), which was the most common bird that has ever existed. Massive flocks literally used to darken the skies of North America when they were migrating, according to contemporary accounts.

At their peak, there may have been more than six billion of these birds. Yet, by 1900, the species was extinct in the wild. When some of their number were shot, the other pigeons did not disperse, though this was not an expression of grief. Instead, their behaviour simply indicated a highly-developed social nature that did not recognise the danger present.

GRIEVING CORVIDS

Only in the case of corvids — members of the crow family — and possibly certain larger parrots is there some evidence that they may react with grief when one of their number dies, for whatever reason.

Professor Marc Bekoff, a biologist from the University of Colorado, recorded the behaviour of a group of black-billed magpies (*Pica hudsonia*) which he observed in the wild near his mountain home, close to the town of Boulder, USA. Almost identical in appearance to their European cousins, these corvids are equally aggressive and opportunistic predators, raiding the nests of other birds to steal eggs or young chicks. They live in loose groups, but, so it would appear, can display signs of grief, revealing a more sensitive side to their nature.

Four of these magpies were observed by Dr Bekoff in the vicinity of a companion that was lying on the ground, having recently been killed by a car. One of the group approached the dead individual, and gently pecked at the body, rather in the way that an elephant will touch the corpse of another, according to Dr Bekoff, and then flew off.

He then noted how another member of the group behaved in a similar fashion, after which one of the birds left, returning moments later with some grass

As members of the corvid family, ravens have been linked with death for centuries, with this belief extending right back to Celtic mythology.

Animal Grief

Magpies have been reported as displaying grief when one of their number is killed.

Sentinel birds may watch for danger within flocks of cockatoos, which means that these birds display signs of social awareness and concern.

in its bill, which it laid gently alongside the deceased magpie. Another bird did exactly the same thing, with all four standing briefly around the body, before finally flying off.

When these observations were first published in the scientific journal *Emotion, Space and Society* in 2009, and then received wider publicity, Dr Bekoff was contacted by other people who claimed to have observed similar ritualistic behaviour not just in magpies, but also other corvids. Magpies are often documented as making a lot of noise when one of their number dies suddenly, perhaps as a result of being shot. This is a good example of why correctly interpreting animal behaviour with regard to death is not an easy or straightforward task: in such instances, these vocalisations probably serve as alarm or warning calls, but it's not hard to see how they could just as easily be interpreted as a display of grief.

There is no doubting the intelligence of these birds, either. In ancient Rome, they used to be taught to talk, and were commonly kept at barbers' shops to amuse waiting customers.

WATCHFUL PARROTS

In the wild, a clear social structure can be observed in a number of species, with pair bonds being maintained between individual birds, even in large flocks. By living in groups in this way, there can be safety in numbers; members of a flock of cockatoos may even take it in turns to watch for danger, allowing the rest of the group to feed in relative safety. If a threat is seen, the sentinels will utter a harsh warning call, alerting their companions.

Should disaster befall one of a pair, though, its partner will not show any overt signs of grief, and, in due course, is likely to form a bond with another member within the flock, although under normal circumstances, depending on the species concerned, parrots will pair for life.

Parrots have also been valued for millennia as companions, with some species being capable of forming a strong bond with people. Outside of the flock environment, individual parrots are more likely to display obvious signs of grief.

Companion animals

Companion animals are a significant part of the daily lives of many people, with typically over half the households in the western world being home to at least one pet. Unsurprisingly, there are many accounts of very close bonds between owners and their animals, and the death of a companion can be exceedingly distressing as a result. But this is no one-way relationship, as there is also clear evidence that animals can, and do, develop strong feelings for their owners, also, and are capable of acting on these feelings.

LOYALTY IN DEATH

Japan is home to a relatively large breed of dog which was originally used by huntsmen tracking dangerous quarry, such as Asiatic black bears. Named after the northern province of Akita where it was developed, the Akita Inu is intelligent and very loyal by nature, forming a strong bond with family members.

On November 10, 1923, an Akita puppy – one of a litter – was born on a farm close to the city of Odate in Akita Prefecture. Hachiko, as he became known, was obtained by Professor Ueno, who worked in the agricultural department at the University of Tokyo. Each day, the young dog accompanied his owner to Shibuya station, returning home afterward,

The Akita Inu has a hardy nature.

and then walking back to the station in the evening to meet the professor's train.

Tragically, during May 1925, the professor suffered a fatal stroke at work, and so never caught the evening train home. Hachiko was alone. Various attempts were made to find him a new owner, but

Animal Grief

he simply escaped and returned to the home he had shared with the professor, and where he had grown up. He also continued to visit the station at the same time every day when the train was due, in the hope of being reunited with Professor Ueno.

Fellow commuters, who had seen Hachiko there in happier times with his owner, now brought him food, and one of the professor's former students helped to publicise his dog's story while researching the history of the breed. At that stage, there were just 30 documented pure-bred Akita Inus in Japan, including Hachiko.

An account of Hachiko's remarkable devotion to his owner then appeared in one of Tokyo's leading newspapers, and evoked an amazing response. His loyalty became a subject of national pride, and Hachiko's story also served to raise awareness of the Akita Inu's plight, helping to save this ancient breed from extinction as a result.

Hachiko continued spending his time on the streets, visiting the station every day, for nearly ten years after his owner's death. When Hachiko himself died on March 8, 1935, his body was placed on display at the National Science Museum of Japan, where it can still be seen today.

At Shibuya station, a statue of Hachiko has become a prominent landmark and a popular meeting area, while the exact spot where he waited is indicated by bronze paw prints. On April 8 each year, hundreds of people still attend a remembrance service for Hachiko, held at the station.

Perhaps unsurprisingly, there have been a number of popular films about Hachiko, as well as documentaries, which demonstrates our fascination with a dog that showed such devotion to its owner. In fact, interest in Hachiko has, if anything, intensified since the turn of the century. Well-known Hollywood actor Richard Gere starred in the movie entitled *Hachiko: a dog's story* which had an international release in 2009, grossing over $50 million worldwide.

Hachiko captured the hearts of the Japanese nation.

Probably one of the most poignant tributes, though, was made possible by technicians working on behalf of a Japanese radio station. They managed to restore an old broken record which featured Hachiko barking. It found millions of listeners when it was broadcast in 1994, some 59 years after his death.

DIFFERENT INTERPRETATIONS

The devotion and loyalty displayed by Hachiko to his owner are not unique, however, and it is certainly unclear whether or not Hachiko understood what had actually occurred. The question of whether this behaviour was essentially just a habit, with Hachikō having regularly undertaken this journey at a set time, back and forth, before the tragedy occurred, also remains. If it was, it could then be argued that the routine was developed further after his owner's death, by what is often known in behavioural terms as 'reinforcement,' people providing food for Hachiko when he was at the station.

However, the one aspect which strongly suggests that Hachiko was missing his owner is that he kept returning to his original home, when attempts were made to move him to various other locations in the city, after his owner had died. While Hachiko may not have appreciated what had actually happened to

Professor Ueno, a very clear case can be made to suggest that he was nevertheless missing the person who was the central figure in his life, and, in effect, grieving for him.

ANOTHER SIMILAR INSTANCE

Hachiko's story is well-known worldwide, but equally, there are other cases like this which are well-documented, although they have not been so widely publicised. A similar instance occurred in Italy, involving a puppy called Fido who was adopted by Carlo Soriani, a man from the town of Luco di Muggello. During the Second World War, on December 30, 1943, Carlo was amongst those who lost their lives in a bombing raid over Borgo San Lorenzo, near Florence, and so he never returned home.

Each day thereafter until his death 14 years later, Fido would continue to visit the bus stop which Carlo had used while travelling back and forth to work, in the hope of being reunited with his owner. Fido was ultimately buried near his owner in Luco's cemetery, and his loyalty is commemorated in a sculpture by Salvatore Cipolla (1933-2006), located outside the town hall.

GRIEF-STRUCK

Perhaps an even more clear-cut display of actual grief by a dog dates back to the nineteenth century, recorded in the Scottish city of Edinburgh. A young Skye Terrier, christened Bobby, was obtained as a puppy by John Gray, who was employed by the local police force. They forged a very close bond, before tragedy struck on February 8, 1858, when Gray died of tuberculosis.

He was buried in a graveyard forming part of Greyfriars Kirkyard, in the old part of the city, and it was here that Bobby took up residence after his owner's death. People marvelled at the dog's devotion. He seemed instinctively to know that John Gray was there, possibly having been able to detect his scent. Especially in his latter years, Bobby was persuaded to

rest in houses near the graveyard, but he never left the area.

His loyalty could have been fatal, as dogs without owners were likely to be put down, but the Lord Provost — the head of the city — paid Bobby's license fee, and the little terrier was fitted with a

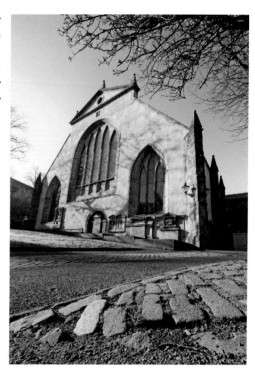

Greyfriars Kirkyard in Edinburgh, Scotland, where Bobby waited for his owner.

distinctive brass collar, with an inscription confirming this fact. Bobby lived to a remarkable age, particularly considering his harsh lifestyle, although local people in the area did provide him with food. He was about 16 years old when he died on January 14, 1872.

Since Greyfriars Kirkyard was consecrated ground, it was not possible for Bobby to be interred next to his owner, but the decision was taken to bury him nearby, inside the gate of the cemetery. Such was the affection that this little dog inspired that a

Animal Grief

variety of monuments have been commissioned to commemorate his memory over the course of more than a century. His grave is now marked by a red granite headstone, paid for by The Dog Aid Society of Scotland in 1981.

There have also been many books and films based on the story of Greyfriars Bobby as he has become known, though not all have been strictly accurate. Yet this little Skye Terrier typified the sense of loyalty that many owners believe that their dogs possess. There was no reason for Greyfriars Bobby to remain in close proximity of the churchyard, as it was not an area that he would have been well acquainted with before his owner's death. Instead, presumably driven by a desire to remain close to John Gray, he adopted this area as his territory, getting as close to his owner as he possibly could.

LOYALTY AFTER DEATH

There are a number of other, similar accounts of canine loyalty on record from around the world, although not all the dogs involved have shown the lifelong devotion of Greyfriars Bobby. Catastrophic mudslides killed over 600 people in January 2011, in the vicinity of Rio de Janeiro in Brazil. Among the victims was Cristina Cesário Maria Santana, and following her burial, people recorded how her dog, Leao, maintained a vigil next to her grave for more than two days.

A not completely dissimilar story from the USA relates to events that took place during August 1936, in the town of Fort Benton in Montana. A casket containing the body of a shepherd was being loaded on to a train heading east, when Shep first showed up at the station. It soon became clear that, almost certainly, the body being sent home for burial was that of Shep's owner.

Shep, a Collie-type dog, then came back repeatedly to the station on a regular, daily basis, hoping in vain that his owner would return, to such an extent that he became a common sight around the

This life–size sculpture of Greyfriars Bobby by William Brodie was commissioned by Baroness Burdett–Coutts, soon after Bobby's death in 1872. It can be seen at the corner of Candlemaker Row and the George IV Bridge in Edinburgh.

station. Further tragedy struck approximately six years later, on January 12, 1942, when Shep slipped under the wheels of a train coming into the station, and died. His loyalty is now commemorated by a bronze sculpture in the town.

STAYING PUT

In the violent encounters linked to land ownership which flared up in the African country of Zimbabwe during the early years of this century, farmer Terry

The statute of Shep, which can be seen in Fort Benton, Montana.

Ford was murdered on land to the west of the capital, Harare. He had three dogs, two Border Collies and an old Jack Russell called Squeak, who was 14 years of age. Squeak had been devoted to Terry throughout his life, and was with him at the time of his death.

When his owner's body was discovered, Squeak was found alongside him, in a very distressed state. He was curled up in the crook of Terry's legs, and it took more than an hour to persuade him to leave. Squeak carried on showing signs of distress for some time afterward. Meanwhile, the two Collies seemed largely unaffected, suggesting that Squeak was clearly grieving for his deceased owner, to whom he was so close.

There are various other accounts of dogs staying with their owners who have died, watching over them. Another case involving a Jack Russell Terrier occurred during 2001 in the Scottish Highlands. The alarm was sounded when a retired newspaper sub-editor called Graham Snell failed to return from a hike in the vicinity of Altnaharra, with his dog, Heidi. The rescue party found Heidi, who was unhurt, alongside Graham's body. He had fallen at least 150m (500ft) to his death, two days previously.

Was it devotion or grief? Once again, the difficulty of coming up with a clear definition of grief means that it is impossible to say with certainty. What is clear is that, once again, the dog had recognised that something was wrong, and had chosen to remain with her owner.

Such loyalty is not a feature seen only in Terriers, though, as similar reports cover a variety of other types of dog. Buddy, for example, was a black Labrador Retriever, kept by Bill Hitchcock on the remote outpost of Knight Island. This island forms part of the Aleutian chain, located off the south-western coast of Alaska. The pair had been living at a remote lodge over the winter, with the temperature there dropping as low as -23°F (-31°C) at that time of year. They had headed off by boat in search of firewood, but Bill was killed by a bough which fell on him.

Buddy appeared to have stayed most of the time with his dead owner over the course of the next 12 days, facing appalling weather conditions on the island, before greeting rescuers when they arrived, and leading them to the spot where Bill lay dead. Over 1000 offers of a new home for brave Buddy flooded in, and he ended up being rehomed with Mayor Jim Brewer from the Alaskan Peninsula village of Chignik.

Unfortunately, however, the story had an even sadder ending. Barely two months after adopting Buddy, the mayor could no longer cope with him, as the dog was repeatedly aggressive. Having had his hand bitten badly, Brewer finally had Buddy euthanased. After living entirely on his own on the island, Buddy clearly could not adapt to these new surroundings, having been devoted to his previous owner.

CANINE GRIEVING

There are also many accounts of dogs grieving for canine companions, typically when they have lived and grown together from puppyhood. The signs of grief in such cases are remarkably consistent, as demonstrated by the case of Sandy, an 11 year old Chihuahua owned by Ann Bell from Tuffley, in the English county of Gloucestershire.

Sandy started to sleep for much longer than

Animal Grief

Having lost an animal companion, an elderly dog like this Golden Retriever can find it difficult to adapt to life alone.

before, lost his appetite, and appeared to be looking for Rosie, his long-term companion who had been four years older when she died. In this case, though, a cardboard cut-out provided by Ann seemed to work, persuading Sandy that Rosie was still there, as he then appeared less depressed and his appetite recovered.

FELINE PERSPECTIVES

Cats, of course, do not have the same reputation for loyalty to their owners, or others of their own kind, when compared with dogs. They are far more solitary creatures by nature, although in many homes, cats can form a surprisingly close bond with dogs, and clearly, they may communicate with each other. Dogs may display signs of grief, rather as if one of their own kind has died, should their feline companion pass away, although cats rarely show any obvious indications of loss in the reverse situation.

Even more surprising, perhaps, than a close association between a cat and a dog, is the relationship that can develop between stable cats and horses. Nervous thoroughbreds in particular seem to tolerate the presence of cats without becoming upset by them, as might be expected.

In fact, the Godolphin Arab — one of the three stallions from which all of today's thoroughbreds are descended — was very attached to a tabby cat called Grimalkin, who lived in the yard where he was kept. After Grimalkin was killed in an accident, however,

Dogs and cats can become very close, especially if they grow up together.

Animal Grief

Close relationships between horses, like this Belgian draft horse, and cats are not unusual.

the horse developed an instinctive and active dislike of all other cats, which remained very apparent right up until his own death in 1753. Perhaps this anger was a sign of grief?

BIRD TALK

Companion birds, too, can react with anger under certain circumstances, particularly in the case of members of the parrot family, which are popular as talking companions. Their ability to talk varies significantly between species, and, just as was the case with Koko the gorilla (see chapter 3) who learnt to sign, the question that has fascinated people is whether parrots can understand the meaning of words, and express their actual feelings in language.

Much of the most informed work in this field of research has been undertaken by Dr Irene Pepperberg, working with her star pupil, a grey parrot (*Psittacus erithacus*) who was called Alex. He was obtained from a pet store in 1977, when he was a year old. During his lifetime, Alex contributed significantly to advancing our understanding of the cognitive

and communicative abilities of this species, before passing away unexpectedly at the age of 31.

It is clear that these parrots at least possess the ability to reason, and do understand the meaning of words. Alex, for example, would refuse an item if it was not what he had requested. Investigations into grief, however, are more observational in the case of these birds, rather than being built on dialogue.

Grey parrots are very sensitive by nature, and will react to stresses in their lives by plucking their feathers, notably the short contour feathers which cover their underparts; in more extreme cases, they will also pull out these feathers in the vicinity of their shoulders and upper back. There is no single cause of this behaviour, but it can often be linked with the death of a mate in the home or in aviary surroundings.

This species is one of many members in this group of birds where there is no clear way of distinguishing the sexes by visual means. Breeders must therefore rely on DNA sexing for this purpose, in order to obtain a breeding pair. But this is known to be only one part of the issue, when it comes to achieving a successful breeding with this species. Mate choice is highly significant: simply placing two adult grey parrots of the opposite sex together is no guarantee that they will prove compatible, and start nesting.

African grey parrots rank as one of the most talented birds, in terms of their ability to talk.

Animal Grief

ignore each other. Clearly, something more than a simple desire to mate is involved here, and it can only really be described in terms of physical attraction.

Under these circumstances, it is therefore not entirely surprising if, when one of the birds dies, especially if they have been together for decades, the survivor grieves for his or her dead partner. This is most likely to be manifested by feather-plucking, although there are other reasons for this behaviour, including an unsuitable diet.

There is no doubt, however, that feather-plucking can become habitual, and is also usually stress-related, given that treatment with drugs such as clomipramine have been used successfully to remedy this problem. This medication, formulated as an anti-depressive for human patients, is known to operate by maintaining the level of neurotransmitters, particularly serotonin, at receptors where they can act, blocking their uptake into the central nervous system.

Boredom and isolation can also play a part in triggering feather-plucking, arising from a change in the bird's daily routine. This will occur when one of a pair of pet parrots dies, leaving the surviving individual alone, bereft of companionship. Such environmental changes can then lead to feather-plucking, which is indicative of grief in these birds. Moving a well-settled pet bird into a new home may have a similar effect. Even if the cause of the problem is identified and corrected, this behaviour can easily become habitual without medical intervention.

Plenty of evidence exists to show that larger members of the parrot family are just as selective in terms of their affections to people as they are to

There are cases where pairs of grey parrots have been together for the best part of a decade, but have never bred, and yet, when introduced to a new prospective partner, the hen may suddenly start laying within a few weeks. Preening between members of a pair is much stronger than would be the case with birds who do not get on, and may even completely

A pair of colourful scarlet macaws reinforce the bond between them by preening each other.

choosing prospective mates, but they can be fickle on occasion. This may reflect the fact that not all parrots form life-long pair bonds, and when faced with separation, they can be opportunistic.

A well-known bird-keeper of the 1920s, W H Workman, kept a white-winged parrakeet (*Brotogeris versicolurus*) which showed a strong dislike for his maid. The bird would screech loudly when he saw her, and would try to bite her on occasion, too. Yet, when Workman went away, leaving a slightly hesitant maid in charge of this feathered fiend, the parakeet became charm personified, displaying no sign of aggression toward her in Workman's absence. On his owner's return, however, the parrakeet fell back into his old ways, actively expressing dislike for the maid at every opportunity.

Animal Grief

It is often said that grief takes many forms, and although cats do not seem to display the same sense of loss as dogs as a result of death, individuals can still show a remarkable devotion to their owners. They can grieve for missing owners, and may, under certain circumstances, endeavour to be reunited with them.

The most famous case of this type involved a cat called Sugar. When Mr and Mrs Woods decided to leave California and relocate to Oklahoma in 1951, they opted to leave Sugar behind with neighbours, as they thought that she would become very upset on the 2400km (1500ml) journey. They were also worried that she might not settle in the unfamiliar terrain once they had arrived.

About a year later, Mrs Woods was outside her new home

A parrakeet of the type which was kept by Workman.

Signs of feather-plucking are evident in this scarlet macaw. Compared with parrots, other groups of birds are far less susceptible to this behaviour.

when a black and white cat jumped on to her shoulder. She was amazed to see that the cat looked just like Sugar; on closer examination, she noticed the deformed hip joint of the cat's left leg. It had to be Sugar. On further investigation, it transpired that Sugar had disappeared from her new home soon after the Woods had left, and had not been seen since.

Another case in England had a less happy outcome, however, when a cat called McCavity decided to return to his former home in Cumbernauld, Scotland. He had recently moved with his owners to the opposite end of the country, to the city of Truro in the south west of England. This marathon trek of some 800km (500ml) back to Scotland took him just over three weeks, but he never recovered from his efforts, and died soon afterward, sadly.

What is it that motivates cats to travel such distances? Is it grief that plays a role, and if so, was Sugar so distraught about being separated from her owners that she decided to set off to find them? While it is possible to find an explanation for McCavity's journey, grieving, in this case, perhaps for familiar surroundings, how could Sugar possibly have tracked down her owners in Oklahoma?

A number of animals such as sea turtles and swallows can travel thousands of kilometres away from their nesting areas, and yet still manage to find their way back there to breed. In the case of turtles, this can even be after an interval of perhaps a decade. Scientists believe that the earth's magnetic field plays a key role in assisting the animals' navigational skills.

Although the majority of feline cases relate to individuals returning to their former haunts, a number concern cats tracking down former owners who have moved into previously unknown areas. While Sugar is an extreme example, there are various other reported cases of cats having been able to successfully track down their former owners, and being reunited with them. Another daunting marathon journey of this type was undertaken by a cat that followed its veterinarian owner right across the USA, from New York to California, settling immediately back in his old familiar chair once arrived.

This ability of cats to set off into the unknown

Cats may travel great distances, possibly influenced by the grief of separation in some cases.

75

Animal Grief

and find a person was first described as psi-trailing by Dr Joseph Rhine (1895-1980), who began his scientific career as a botantist, but subsequently developed a fascination with parapsychology. While based at Duke University in North Carolina, he studied a number of cases of this type.

Rhine's investigations revealed that it is not just cats that can display this ability of psychic-trailing, documenting the case of a pigeon that had been nursed back to health by a young boy. When the boy was taken into hospital, the pigeon suddenly appeared unexpectedly outside the window of the hospital room where he was staying. These are known to be birds with strong homing instincts, but again, this was another case which defied obvious explanation.

RADICAL THOUGHTS

Could it be that, through grief or the trauma of separation, a psychic connection can be established between people and their animals, which we do not currently recognise? It would be easy to dismiss this notion as fanciful, but in one group of mammals at least, the evidence points toward telepathic ability. This

Pigeons are well–known for their homing instincts.

Telepathy is now believed to play a part in communication between dolphins.

Like us, birds such as parrots rely very heavily on their sense of sight, and this perhaps can help birds pick up on visual clues from their owner.

Animal Grief

information has been obtained by serious laboratory research, carried out by scientists studying dolphins on behalf of the US Navy, and others working for various universities.

It tends to be those species which are capable of expressing grief **that are also** linked with telepathy. Many owners, of course, claim that their companion understands what they are thinking. Aimée Morgana believed that her African grey parrot, N'kisi, was able to pick up on her thoughts and elucidate them, thanks to his ability to speak, combined with a vocabulary of over 1200 words. Tests in a controlled environment, with N'kisi totally separated from his owner, did indeed suggest statistically that the bird could pick up on his owner's thoughts. But could he detect her emotions, too, and respond to them? This is something that simply cannot be explained at present ...

DIFFICULTIES IN THIS FIELD OF STUDY

It can be easy to get carried away by stories of animal grief, though, in the case of domestic pets, because human nature is such that we want to empathise and identify with them, and have them empathise with us, bringing us closer together. Stories in this area can easily develop on this basis, even though the underlying explanation is actually quite different from how the situation may initially appear.

Take the case of the cat lying on this tombstone. As it does so regularly, it would be tempting to suggest that the cat may have been linked to the person buried here. In reality, though, what is happening is that, in the morning, with the grass damp from the overnight dew, the cat has sought a more comfortable place to sunbathe. This particular grave enables the cat to stretch out, and even when the sun is not particularly strong, the surface attracts and holds heat, providing a warm place to lie.

IN CONCLUSION

Just because certain cases can be discounted, however, does not mean that animals are incapable of grieving. Quite the contrary. There is now indisputable evidence that this emotional awareness and capability exists in a number of species. As with people, though, the way in which grief manifests itself varies, with much still remaining to be discovered about this new field of animal behaviour.

So, what actually is animal grief? It is perhaps best defined as a heightened emotional state of awareness that affects an individual's behaviour so that it reacts in a different way to what would normally be anticipated. It is a multi-faceted and even a pragmatic emotion, rather than a passive sensation. It can reinforce survival instincts, binding groups together in the face of adversity. With companion animals especially, it is not just death but separation, too, that can lead to grief.

After being in denial for centuries, recent discoveries mean that we as a species can finally start to empathise with some of our fellow creatures under such circumstances.

Myths can arise about animal grief by misinterpretation of animal behaviour, making this a potentially difficult field to investigate and research.

INDEX